本书效果图

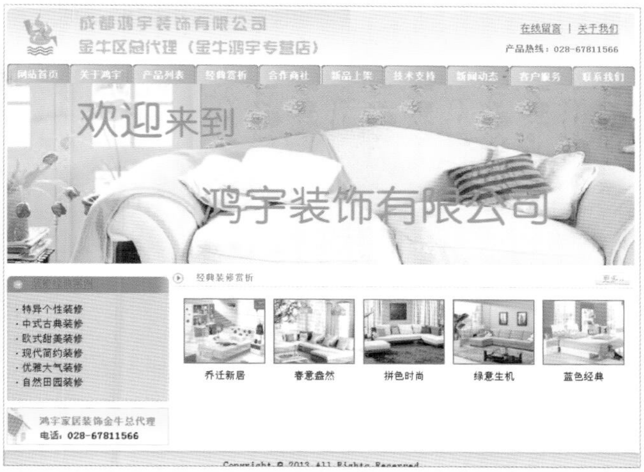
成都鸿宇装饰有限公司
金牛区总代理（金牛鸿宇专营店）
在线留言 | 关于我们
产品热线：028-67811566
网站首页
关于鸿宇
产品列表
经典赏析
合作商社
新品上架
技术支持
新闻动态
客户服务
联系我们
欢迎来到
鸿宇装饰有限公司
经典装修赏析
·特异个性装修
·中式古典装修
·欧式甜美装修
·现代简约装修
·优雅大气装修
·自然田园装修
乔迁新居
春意盎然
拼色时尚
绿意生机
蓝色经典
鸿宇家居装饰金牛总代理
电话：028-67811566

Wedding摄影

欢迎来到MY ART网站……
我的艺术格言……
作品展示

彩霞
CaiXia Style Webset
主页
个人简介
我的生活
我的一家
留言
个人介绍

Play
Site
主页
音乐
电影
电视剧
娱乐

本书效果图

free time
奶茶
特色冷饮

free time
奶茶
分店信息

我的一家
My Family

摄影之家
IMAGE picturesque

Category 1
Category 2
Category 3
Sed ut perspiciatis
Unde omnis iste
Natus error sit volup
Tatem accusantium

本书效果图

home
about
articles
contacts
site map
MUSIC BEATS
FEEL THE RHYTHM OF THE NIGHT
Latest News!
Burn The Night!
Recent Articles
About This Template
Music Beats
Night Club Life
Best DJ Sets
Copyright - Type in your name here

摄影之家
你的生活记录之旅
在这里 登录
加入摄影之家
用户留言
常见问题
PHOTOGRAPHY
摄影之家&记录

Bridal Shop

开始游戏

MUSIC

本书效果图

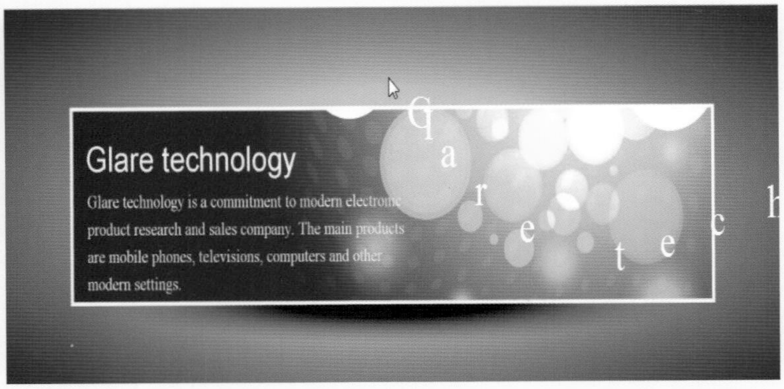
Glare technology
Glare technology is a commitment to modern electronic product research and sales company. The main products are mobile phones, televisions, computers and other modern settings.

清雅飘逸

Smile Cafe

用户名 abs113700
密 码
登 录

美人如画

only lady

Take out TV!
DMB PHONE
HIT ZONE

HAOJIAJU 好佳具
四川好佳具有限责任公司

PHOTO STUDIO
World Expo 2009 festival

겟한방원
진료안내
온라인상담
온라인예약
당신의 진정한
아름다움을 되찾다...
01 한방클리닉
맞춤 한약재보기
02-3446-8621~3
겟한방원

HAPPY VIRUS
THE FREEGINE FOUNDATION
나눔으로 다시 푸는
인생수업
행복한약속

ASA web
COMPANY
PRODUCT
CUSTOMER
RECRUIT
COMMUNITY
DIGITAL DREAM UTOPIA
BEST PRODUCTS
ASANOTE 705
BUSINESS INTRO
NEW PRODUCT

ASAWEB
COMPANY
PRODUCT
COSTOMER
RECRUIT
COMMUNITY
당신의 비즈니스 파트너
NO.1
BUSINESS
NOTICE
NEW PRODUCT
WEB SERVICE
ASAWEB

SaDing PIZZA
Welcome to Sading
营业时间: 星期一~星期五: 10am - 10pm / 星期六~星期天: 9am - 11pm
餐厅地址:
最新美食
特色美食集锦
评论/留言
John Doe
联系电话: +028-6874-316

GREENY
照片精选
分类查看
订阅

Siu School
ACADEMY
Welcome to Our School
信息中心
意见调查
8665998

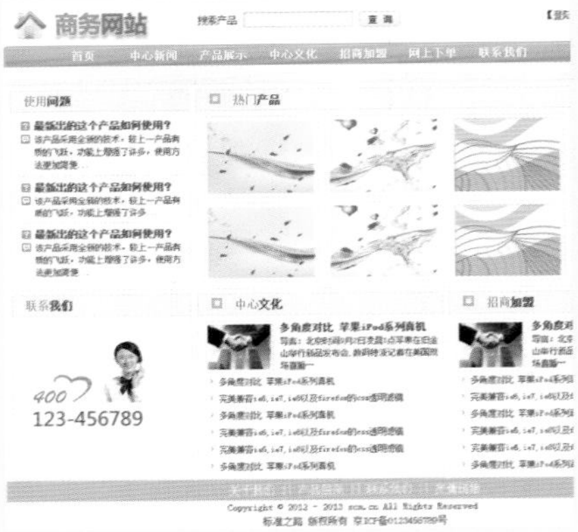
商务网站
使用问题
热门产品
联系我们
中心文化
招商加盟
400
123-456789

Delicious Western
TH ME WESTERNT
DELIVERY NOW!!!

Qiate Plant
Plant Navigation

About restaurant
关于餐厅